USBORNE

HELPING OUR PLANET

Expert advice from

Paul Brown

Former Environment Correspondent

for The Guardian and Co-founder

of Climate News Network

USBORNE

HELPING OUR PLANET

JANE BINGHAM

Designed by Kirsty Tizzard

Illustrated by Sara Rojo,
Christyan Fox, Nancy Leschnikoff
& Freya Harrison

Edited by Felicity Brooks & Jenny Tyler

INTRODUCTION

There's no doubt about it — our planet is in trouble. But what can **YOU** do to help? This book is full of practical tips and advice to help you change your behavior, influence the people around you, and get your message heard.

You'll find chapters on planet-friendly shopping, eating and traveling, and on ways to save energy and cut down on waste. There's also advice on getting drastic about plastic and taking better care of the natural world.

Caring for the Earth is a massive challenge facing us all today, and every single one of us needs to play our part. We must work together to save our amazing planet — and we can all get started **NOW!**

CONTENTS

Continued over the page

TAKE
ACTION
NOW!

USBORNE QUICKLINKS

There's lots of information on the internet about helping our planet, but it's important to know which sites you can trust. At Usborne Quicklinks, you can find practical guides to planet-friendly activities, from creative upcycling to growing your own vegetables. You'll also find sites filled with facts and figures on major issues, such as plastic pollution and climate change.

For links to all these sites, go to:
usborne.com/Quicklinks
and type in the title of this book.

When using the internet, please follow the safety guidelines shown on the Usborne Quicklinks website. Children should be supervised online.

1

WHAT'S HAPPENING TO OUR PLANET?

The Earth has been home to humans for millions of years, providing us with everything we need for life. But now our planet is in trouble ...

The Earth's not looking too good.

Rising pollution
Climate change
Plastic waste
Declining wildlife
Disappearing forests
Fires and floods
Global pandemics

These are worrying symptoms.

We must help our planet recover.

HOW DID IT HAPPEN?

Sadly, it's we humans who are to blame for the state of our planet. Around 250 years ago, humans invented machines that needed power to make them work. People burned coal, oil and gas (known as fossil fuels) to provide the power for their factories and engines, and this burning process sent out harmful gases into the atmosphere.

Over the next two centuries, people across the world came to rely on fossil fuels to power their industry, transportation, lighting and heating. This created a build-up of gases in the atmosphere, which has led to dangerous pollution and worrying changes in our climate.

You'll learn about these problems later in the book AND find practical ways to help tackle them.

WHAT ABOUT OTHER PROBLEMS?

Humans have been responsible for other problems too...

As we produce more and more stuff, we are piling up enormous amounts of waste. **Pollution** from human trash threatens the health of humans, animals and plants. And plastic waste is the biggest threat of all, because it doesn't rot away.

Over the centuries, humans have destroyed many **natural habitats**, putting animals and plants at risk. Now this process is really speeding up, with over a million species in danger of **dying out**.

Large areas of **rainforest** have been cut down and **large-scale farming** practices have added to the problems of pollution and climate change. Meanwhile, the combination of human actions and climate change have led to **severe weather**, **fires** and **floods**.

WHAT'S BEING DONE TO HELP?

Sometimes, the news of our planet's health can seem very bleak, but there is GOOD NEWS too. All over the world, people are waking up to the Earth's problems and trying to find new solutions ...

Here are just a few of the POSITIVE MOVES that people are making:

Scientists are developing alternative energy sources to replace fossil fuels.

Governments are introducing laws to reduce pollution and cut down on waste.

Conservationists are working on ways to preserve wildlife.

Companies are creating products that cause little harm to the Earth.

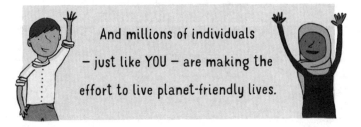

And millions of individuals – just like YOU – are making the effort to live planet-friendly lives.

SO, WHAT CAN <u>YOU</u> DO TO HELP?

This book covers a range of ways that you can make a difference to the health of our planet. Look out for guidelines on all these topics:

→ Saving energy

→ Reducing waste

→ Cutting down on plastic

→ Planet-friendly shopping, eating and travel

→ Looking after our water

→ Caring for the natural world

→ Spreading the message — and getting yourself heard

You'll find helpful suggestions, ideas for projects and lots of practical tips to follow.

Let's ALL join in the effort to take care of our world!

HOW PLANET-FRIENDLY ARE YOU?

2

Are the things you do every day helping or harming our planet?

The quick quiz on the next page is designed to get you thinking about your daily habits. Choose the answers that are closest to your behavior, and don't be too worried about getting them "right." The aim is simply to make you more aware of your actions and choices and their possible impact on the planet.

QUICK QUIZ

Which of these answers – A, B or C – best describes you and your daily life?

1. When you're taking a shower, do you:
    A. Make sure you're finished in 3 minutes.
    B. Spend less than 10 minutes in the shower.
    C. Need at least 15 minutes in the shower before you're ready to face the day.

2. If you're feeling thirsty, do you:
    A. Fill up the water bottle you always carry with you.
    B. Buy a bottle of water and reuse or recycle it later.
    C. Buy a bottle of water and chuck it when it's empty.

3. In an average week, do you:

    A. Usually walk or cycle.

    B. Mainly travel by bus or train.

    C. Make most of your journeys by car.

4. If you're choosing a sandwich or wrap for lunch, do you:

    A. Pick a meat-free filling.

    B. Choose fish or chicken.

    C. Go for something seriously meaty, such as beef or ham.

5. If you're feeling cold at home, do you:

A. Put on an extra layer and move around some more.

B. Turn up the heating a few degrees.

C. Turn up the heating high — you hate feeling cold!

6. When you're opening a present, do you:

A. Open it very carefully so you can use the wrapping paper again.

B. Rip off the wrapping and stick it in the recycling bin.

C. Chuck all the wrapping into the general trash — you're too excited to think about saving paper right now!

ANSWERS

Mostly "A" answers

You're an eco warrior, but you can still benefit from the tips in this book.

Mostly "B" answers

You want to help the planet, but you could work on being a bit more planet-friendly.

Mostly "C" answers

You'll have to change your habits if you want to give your planet the help it needs. But don't worry — reading this book is a great place to start.

3.

THE BIG ENERGY CHALLENGE

We rely on energy for everything we do. We simply can't do without it — and we're using more and more of it every year.

How would I ever get started without electricity?

Of course, it's a wonderful thing to have a constant energy supply, but we're starting to realize we can't take it for granted. There are some BIG problems with our power sources.

SO, WHAT ARE THE PROBLEMS?

Most of the world's energy comes from coal, oil and gas. These are known as "fossil fuels" because they formed from the remains of plants and animals that were buried underground millions of years ago. But fossil fuels cause serious damage to our planet:

- Fossil fuels have to be burned to release their energy and this process has some <u>very dangerous</u> side effects.

- Getting the fuels out of the Earth also causes <u>lasting harm</u> to the environment.

You can discover more about these problems on the next few pages. You'll also learn about some <u>solutions</u> that people are finding, as we all face up to ...

THE BIG ENERGY CHALLENGE

A BURNING ISSUE

Oil, coal and gas are burned in power stations to release their energy and produce electricity. They're also burned in factories, cars, planes and trains and by furnaces in people's homes. All this burning sends out **harmful gases** into the air.

The gases produced by fossil fuels are often called **carbon emissions** because they include large quantities of carbon dioxide (CO_2). They are also known as **greenhouse gases**.

WHAT ARE GREENHOUSE GASES?

The term "greenhouse gases" describes a group of gases that build up in the Earth's lower atmosphere. The gases include **methane** and **nitrous oxide**, but the main greenhouse gas is **carbon dioxide**. Methane is given off by farm animals that belch and pass gas, and by rotting trash, while nitrous oxide is released from farm soil as well as from burning fossil fuels.

THE GREENHOUSE EFFECT

As greenhouse gases build up, they act like the glass in a greenhouse, trapping the Sun's warmth and causing temperatures on Earth to rise. This is known as the **greenhouse effect** and it is the cause of **CLIMATE CHANGE** – the biggest threat facing our planet today.

MINING DANGER

Fossil fuels cause problems when they're burned AND when they're extracted from the ground. Mining and drilling damage natural habitats and can lead to dangerous earth movements. Unless these processes are very carefully managed, harmful chemicals can be released into the Earth's underground water systems, causing polluted water to enter rivers and seas.

WHAT ABOUT FRACKING?

Some people claim that fracking is a "cleaner" way to extract fossil fuels from the Earth. In the fracking process, large amounts of water, sand and chemicals are injected into rocks to release trapped gas. Fracking creates fewer carbon emissions than processing coal or oil, but it can still cause serious problems.

FRACKING PROBLEMS

Fracking relies on water — a natural resource that's in short supply in many places. Some of the chemicals used in fracking, such as arsenic, are highly toxic. The process also releases methane gas, which may leak into the atmosphere and contribute to climate change.

There is also evidence that fracking can cause earth movements. For all these reasons, many people are opposed to it. And fracking still produces fossil fuels with all their problems.

Extracting more fossil fuels is NOT the answer.

We need to find ALTERNATIVE energy sources.

WHAT ARE THE ALTERNATIVES?

On the next few pages you can see a range of alternatives to fossil fuels. Some, such as **nuclear** and **hydroelectric power**, have been used for years. Others, such as **solar** and **wind power**, are part of a recent move to use "green" energy sources that do less damage to the Earth.

HYDROELECTRIC POWER

Hydroelectric power is produced in countries with heavy rainfall. Water is collected behind a massive dam and released at very high pressure. The water turns a set of giant wheels, known as turbines, which convert the power of the water into electricity.

But building dams means flooding large areas of land — bad news for the people and wildlife living there.

NUCLEAR POWER

Nuclear power is made from the metal uranium, using a process known as nuclear fission. This process doesn't release waste gases, but it does produce **radioactive waste**, which can be lethal to living things and remains a threat to life for <u>thousands</u> of years. Nuclear waste needs to be stored VERY securely in order to prevent hazardous leaks.

Mining uranium is an expensive process that damages the Earth...

...and many people are opposed to nuclear power on safety grounds.

SOLAR POWER

Solar power is produced by converting heat and light from the Sun into electricity. There are two main methods of producing solar energy: one using solar panels and one using huge mirrors.

Solar panels

Solar panels are often known as photovoltaic or PV panels. They contain light-sensitive cells that convert the Sun's energy into electricity, and they can be mounted on buildings or set up in rows on solar farms. The cells work even on cloudy days, but they produce more electricity when the Sun is shining brightly.

Solar mirrors

Solar mirror farms are set up in deserts and very hot places. Rows of mirror panels the size of garage doors reflect rays from the Sun and use them to heat liquids that boil at higher temperatures than water. These superheated liquids turn water into steam which drives a set of turbines to generate electricity. This type of energy is known as Concentrated Solar Power or CSP.

Every day, the Sun gives off far more energy than we need to power EVERYTHING on Earth...

So solar energy will NEVER run out – and we can produce it with very little harm to the planet.

WIND POWER

The power of the wind can be converted into electricity by wind turbines — tall machines with huge blades that turn very fast when the wind blows through them. The turning blades of the turbines drive a generator that converts movement into electricity. Wind turbines are often grouped together in giant "wind farms" on hills or out at sea.

One large wind turbine
in a windy place can create enough
electricity to provide power for
1,000 homes.

GREEN HYDROGEN

"Green hydrogen" is the name of a gas that's made by using energy from a "green" energy source, such as solar or wind power. It is produced by a process called electrolysis, in which electricity is passed through water to split it into oxygen and hydrogen. The captured hydrogen gas can then be stored in tanks to be used whenever it's needed. At present, it is very expensive to produce, but scientists predict it will be an important energy source in the future.

GEOTHERMAL POWER

Geothermal power uses heat from inside the Earth to produce energy. Water is passed through hot rocks to create steam, which drives a set of turbines that generate electricity.

Iceland, New Zealand, Japan and the US all have major geothermal power stations and there are smaller projects in many countries across the world.

BIOMASS ENERGY

Biomass energy is often simply called "biofuel" or "biogas." It's made by converting plant or animal material (known as biomass) into electricity, gas or liquid fuel.

Rotting plants, food waste and animal manure all release methane, and this **biogas** can be collected and burned to generate electricity or heat water. Some crops, such as sugar beet and corn, can be converted into liquid **biofuel**, which can be used in cars instead of gasoline.

So – my poop can be turned into electricity! Who'd have thought it?

Biomass energy is usually classed as a "**renewable**" energy source because the materials used to create it can be replaced by new ones.

TIDAL POWER

Some scientists are working on projects that
use the power of ocean tides to create electricity.
Tidal power is not yet widely used, but
it may provide a valuable energy
source in the future.

RUN OF THE RIVER POWER

Run of the river (ROR) projects use the force of
running river water to generate electricity. They
work in the same way as hydroelectric dams, but
cause much less damage to the environment because
they don't need massive reservoirs or dams.

Most ROR plants can't store energy...

...but they can add a useful
extra source of power.

A RANGE OF RENEWABLES

All the energy types shown below are known as "renewables" because they're made from sources that won't run out. They can also be produced with much less harm to the environment than fossil fuels. Other names for renewables are "clean energy," "green energy" and "low-carbon energy."

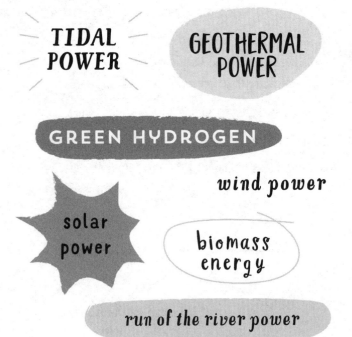

TIDAL POWER

GEOTHERMAL POWER

GREEN HYDROGEN

wind power

solar power

biomass energy

run of the river power

WHAT'S THE FUTURE FOR RENEWABLES?

At present, renewables provide around **one sixth**
of the world's total energy:

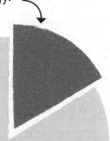

...with fossil fuels
supplying **all** the rest.

This may not sound like much, but the figure is rising
fast. There have also been big advances in storing
energy, so power from renewables can be available
whenever it's needed.

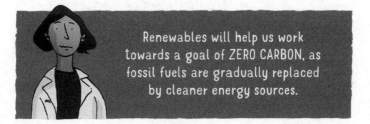

Renewables will help us work
towards a goal of ZERO CARBON, as
fossil fuels are gradually replaced
by cleaner energy sources.

CUTTING CARBON EMISSIONS

Although we're making good progress on renewables, most of our energy still comes from fossil fuels. And using fossil fuels means creating the carbon emissions that are helping to cause climate change. If we want to help our planet, we need to take URGENT measures to reduce our use of fossil fuels.

The MOST IMPORTANT way to cut down on fossil fuels...

...is to REDUCE the amount of energy we use.

A CHALLENGE FOR US ALL

Cutting carbon emissions is a challenge for everyone because we ALL use fossil fuels and create carbon emissions. The impact of your emissions on the Earth is sometimes known as your **carbon footprint.**

LOOKING AT CARBON FOOTPRINTS

Your carbon footprint — very simply put — is the amount of carbon dioxide that you send into the atmosphere as a result of your daily activities. There are websites where you can find the carbon footprint of everyday things we use, including cars, food and clothing. Go to Usborne Quicklinks (see page 9) to find out more.

Having some idea of the size of your carbon footprint can help you recognize how much energy YOU consume. Now is the time to use less energy and reduce your footprint. Are you up to the challenge?

You can follow the tips in this book to help reduce your footprint.

TIME FOR ACTION

So, how can you play your part in the great energy challenge, reduce your carbon footprint, and help our planet? The most important step is to ...

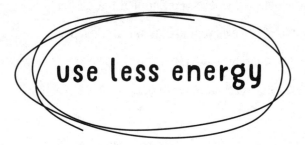

use less energy

Think about all the ways you use up electricity, gas or fuel, and make a BIG EFFORT to use less. You'll find that energy-saving soon becomes a habit — and you can encourage others to follow your lead.

Are you doing ALL YOU CAN to save energy?

TEN SIMPLE ENERGY-SAVING MOVES

1 **Do more things that <u>don't</u> use up lots of fossil fuels.**

Walking, cycling, reading and hanging out with friends are great ways to enjoy yourself while using little or no fossil fuels.

Pedal power rules!

2 **Turn off the lights!**

Get into the habit of turning off the lights each time you leave a room, unless you're coming back in less than five minutes.

Unplug your TV and computer overnight.

Remember to switch off appliances at the wall at the end of the day. In an average home, more than 10 percent of the total household energy is used to power appliances left on standby mode.

OVERCHARGED

Do you leave your phone charging overnight? Most phones are fully charged within two hours. After that, you're wasting electricity for the rest of the night.

4 Don't turn up the heating.

Next time you feel chilly at home, put on more layers and get up and get moving. You could also talk to your family about how your home is heated. Is the heating on when everybody's out? And is your home properly insulated? Remember to keep doors and windows closed to stop heat from escaping.

Moving around soon warms me up!

 Sometimes the problem is feeling too hot! If you have air conditioning, use it sparingly and close your windows or blinds to keep the cool air in.

Dry your clothes the natural way.

Instead of stuffing your clothes into the dryer,
hang them up to dry on a washing line or
clothes rack. This will save you masses of
energy — most dryer models use more energy
than any other domestic appliance.

 Save refrigerator energy.

Wait for hot food to cool before putting it in
the refrigerator, and don't leave the door open
for any longer than you have to. Each minute
you spend deciding what to eat, your
refrigerator is heating up, and it will need
an extra energy boost to
cool it down again.

Help! I'm losing
my cool!

7 **Think before you run your washing machine or dishwasher.**

Always wait for a full load and use a cool or "eco" setting whenever you can. That way you'll save energy and water too.

8 **Save energy while you cook.**

Look at Chapter 8 for ideas on saving energy in the kitchen.

KETTLE CARE

Next time you make yourself a hot drink, remember to boil <u>only</u> the water you need. It has been estimated that the average US family may be using more than twice the energy actually needed to boil the water for their coffee and tea.

9 Look out for chances to use renewable energy.

Buy a battery charger and rechargeable batteries, and go for solar-powered devices, such as lanterns, flashlights and fans.

10 Make a big effort to save energy on travel.

This means thinking about your own travel choices <u>and</u> the transportation involved in the things you buy. You'll find out more about travel in Chapter 9.

MAKING FAMILY CHOICES

It isn't always easy to ask your family to change their ways, but if they see you making changes in your life, they should feel more willing to save energy too.

You can remind the bill-payers in your family that saving energy means **saving money**.

Encourage your family to switch to a "green" energy provider and buy energy-efficient appliances. You may even persuade them to make some bigger moves, such as installing solar panels or changing to an electric car.

Try sharing **facts about fossil fuels** to encourage your family to save more energy. Go to Usborne Quicklinks (see page 9) for links to helpful websites.

MAKING WIDER CHANGES

So, you're doing your best to save energy, but is there more you can do to tackle the energy crisis? Here are some actions that can help to bring about change:

I write to politicians urging them to support energy-saving policies.

I've persuaded some local businesses to switch off their lights overnight.

I take part in marches to raise awareness of climate change.

I've joined a "green team" at my school – looking out for ways to save energy.

Of course, you'll need to decide which actions feel right for you, but one thing is certain: each time you reduce your energy use – or persuade others to be more energy-aware – you'll be helping to preserve our beautiful planet.

4. REDUCING WASTE

Have you ever stopped to think about all the things you throw away?

As well as the everyday trash — the remains of meals, snacks and drinks, the packaging, cartons and bottles, and piles of paper — there's all the stuff we no longer want, such as clothes and shoes, toys, computers and games. The list of things we chuck away goes on and on, leaving us a very big question...

WHAT ON EARTH do we do with all this trash?

MAKING A WASTE MOUNTAIN

If you multiply the things you throw away by all the people living on our planet, it's clear we're creating a MASSIVE amount of waste — and it's getting BIGGER every year. Unless we completely change the way we think about trash, global waste will triple by 2100.

People living in the USA produce enough trash in a **single day** to fill **63,000** garbage trucks. Just think how much that adds up to in a year!

POLLUTION PROBLEMS

Our trash doesn't just create a horrible mess, it also poses a threat to the health of the planet. If we don't dispose of our waste very carefully, the things we throw away can end up polluting our soil and water and harming animals and plants.

WHAT HAPPENS TO OUR TRASH?

After recycling, most remaining household waste is taken to landfill sites, where it is buried underground. But even when trash is buried, these giant dumps still create ENORMOUS problems for the planet.

- Some kinds of trash, such as batteries and plastics, release **poisonous substances** which can escape into the air or leak into the soil. There are strict laws about containing trash in landfill sites, but some people worry that it may not be possible to avoid some pollution.

- Trash is very bulky and more of it is added to landfills all the time — creating the urgent problem of **finding new land** for sites.

- Even trash that rots away naturally harms the environment. For example, rotting food produces **methane gas**, which smells very nasty and contributes to climate change.

- There are also the problems caused by transporting trash to landfill sites — journeys that create harmful **carbon emissions** and air pollution.

WHAT ABOUT BURNING?

Around 12 percent of US household waste is burned and turned into ash before being buried in landfills. Burning trash in an incinerator quickly reduces its bulk, but it has several drawbacks:

→ The burning process needs to be very carefully controlled to avoid pumping out **harmful fumes**.

→ Burning waste **uses up energy** and contributes to heating up the planet.

→ Incinerators produce **toxic ash** that needs to be carefully buried in landfills.

WHAT'S BEING DONE TO TACKLE WASTE?

In the 21st century, we're finally taking steps to tackle the trash problem. Governments are working on safer and more efficient ways to dispose of waste, and scientists are developing a range of planet-friendly products. Some of these products can be recycled and some are made from recycled materials.

Important progress is being made, but we all need to play our part in reducing the trash mountain:

FOUR IMPORTANT WAYS TO CUT DOWN ON WASTE

1 Check what you chuck.

Each time you go to throw something away, remember to ask yourself: "Is it truly trash, or can it be repaired, reused or recycled?" Chapter 5 is full of ideas for making the most of your stuff, reusing and recycling.

Surely you're not going to throw that away?

2 Start sharing and borrowing.

Instead of buying new games, books or magazines, set up a sharing group with your friends. And use your local library to borrow books you don't want to keep forever.

Give it away, don't throw it away.

Even if you see your stuff as a load of old junk, there's still a very good chance that someone could make use of it.

★ Charity stores welcome donations, and it's great to know that the things you no longer want can earn money for a good cause.

★ You could start a bring-and-share club or donate your things to a "freecycle" program.

★ If you have magazines you don't want any more, you could ask your local doctor, dentist or vet if they would like to put them in their waiting room.

4

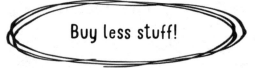
Buy less stuff!

Of course, the best way of all to cut down on waste is simply NOT to buy things in the first place. Kicking the shopping habit will free up your time for other activities. And buying fewer things will leave you with money to spend on interesting experiences rather than extra stuff.

LESS STUFF, MORE ACTION!

Instead of buying presents for friends and family, think about <u>doing</u> something for them instead. Maybe you could cook a special meal or organize an outing or a picnic in the park?

REDUCING PAPER WASTE

Did you know that paper and cardboard make up almost a quarter of our household trash? Fortunately, paper products can be recycled, but recycling still uses up energy.

> Using paper means using up trees. Each year, an area of forest the size of the Netherlands is cut down to make paper products.

If we really want to help our planet, we need to change our attitude to paper. You'll find some helpful ideas for reducing your paper mountain on the next few pages.

SAVE THE TREES!

Trees play a vital role in maintaining the health of the planet. They **absorb carbon dioxide**, helping to reduce the amount of greenhouse gases in the atmosphere. They stabilize the soil, and they provide a home for wildlife. For all these reasons, it's essential to preserve as many trees as possible, especially as our rainforests are being destroyed at an alarming rate (see page 181). If you want to reduce the number of trees being cut down for paper:

➜ Buy paper products made from **recycled paper pulp**.

➜ Buy paper made from wood from a **sustainable forest**.

Sustainable forests are made up of fast- growing trees, so trees that are cut down can easily be replaced. Forest managers promise to plant more trees than they cut down and undertake to protect the wildlife in their forests.

FOUR
PAPER-SAVING TIPS

You can help save trees by buying recycled or sustainable paper products, but it's still essential to reduce your paper use. Try these paper-saving ideas:

 Aim to avoid paper and cardboard packaging – buy loose items whenever you can and use your own bags to carry them home.

 Ask for e-receipts and e-tickets, rather than paper ones.

 Use your phone for shopping lists and notes, instead of a pen and paper.

 Write messages at home on a chalkboard.

CALL JACK!
Maisy-6pm

FOUR IDEAS FOR MAKING THE MOST OF PAPER

1 Write, print and photocopy on both sides of the paper, but always ask yourself:

Do I really need to print this?

2 Turn birthday and Christmas cards into gift tags – or make new cards from them.

3 Recycle envelopes. You can stick a label over the original address.

4 Create your own wrapping paper, using magazine or newspaper pages.

WHAT ABOUT FOOD WASTE?

It's a shocking fact that over a third of all the food we produce globally is thrown away.

We all need to stop taking food for granted and reduce the amount of food we waste.

➡ Aim to shop and cook less wastefully, and keep your leftovers to use the next day.

You'll find ideas for making the most of your food on pages 148 to 150.

➡ If you're eating out, check the portion sizes. If they look too large, share one between two people, or order a smaller portion. And ask to take any leftovers home.

STOPPING "THROWAWAY FASHION"

It's easy to get carried away by the latest wave
of "must have" clothes. But have you ever stopped
to think how throwaway fashion harms our planet?
The process of making clothes creates **pollution**
and consumes huge amounts of **energy** and **water**.
And buying new clothes means throwing out the
old — resulting in more **waste**.

I need to
have this NOW!

But you've
already got 17
tops at home!

New in!

DON'T THROW THEM OUT!

Chucking clothes straight into the trash causes **serious damage** to the planet.

★ Synthetic fabrics, such as acrylic and polyester, contain **microplastics** that don't rot away and can harm animals and plants. (You can read more about microplastics on pages 108 to 111.)

★ Even natural materials, such as wool and cotton, can cause problems for the environment. As they rot, they give off **methane gas**, which contributes to climate change.

The fashion industry is one of the world's biggest waste creators, with a staggering **10,000 items** of clothing being sent to landfill sites **every five minutes**.

SO, WHAT'S BEING DONE ABOUT FASHION?

There are some serious moves in the fashion industry towards more planet-friendly clothing. Some designers are switching from synthetics to natural fabrics. Some are experimenting with compostable fabrics, and some are using recycled textiles in their designs. There are also moves to avoid using plastic and to find alternatives to plastic sequins.

FIVE NEW WAYS TO THINK ABOUT YOUR CLOTHES

 1 Instead of buying heaps of throwaway clothes, go for a few good items and make them last.

2 Try restyling your old clothes to make something new.

 3 Find second-hand bargains at charity stores, thrift stores and second-hand sites online.

 4 If you're buying new clothes, avoid synthetics and choose natural fabrics. Bamboo and hemp are especially planet-friendly because they're made from fast-growing plants that don't require lots of artificial fertilizers.

5 And if you're tired of an outfit, don't throw it away. Give it to a friend or relative, take it to a charity store, or find a way to sell it.

RECYCLING CLOTHES AND TEXTILES

If you're <u>really certain</u> you can't find a use for your clothes, make sure you recycle, rather than throwing them away. All recycled textiles are sorted carefully. Some are sent to countries where clothing is needed, some are used to make paper, and some are used as "industrial rags" for stuffing car seats.

You can recycle your old clothes in a **textiles recycling bank**. Or look out for retailers that **reward you** for recycling your old clothing.

Goodbye old scarf. Enjoy your new life!

CLOTHES BANK

FOUR FABRIC-RESCUE PLANS

1. Stitch panels together to make a patchwork quilt.

2. CREATE COLORFUL FLAGS FROM FABRIC SCRAPS.

3. USE TEXTILES FOR WRAPPING PRESENTS OR COVERING GLASS JAR LIDS.

4. Reuse old fabrics to make dolls and toys.

CUTTING DOWN ON PLASTIC

Cutting down on plastic is so important, there's a whole chapter on it later in this book. But a good place to start is with a waste-busting kit.

MAKE A WASTE-BUSTING KIT

This simple kit is easy to put together and will help save waste every day:

Made from bamboo

Hot drink mug Use it for all your takeout drinks and say "NO" to throwaway cups and lids.

Reusable silverware set Bring it out each time you're offered plastic silverware.

Made from metal

Water bottle Fill it up at home, then top it up as you go.

Food container with lid Use it to carry all your meals and snacks. And if you're buying sandwiches or other takeout items, ask the retailer to use your container.

Shopping bags

Always carry one with you, so you never have to say "yes" to plastic bags.

Cloth napkins

Use cloth napkins instead of paper ones, and wash them after you've used them.

> I take my waste-busting kit everywhere I go and clean it out at the end of the day.

Taking the NO SHOP CHALLENGE means that you STOP buying anything other than <u>essentials.</u> The challenge can last for a week, a month or longer, and you can do it alone or with friends.

The NO SHOP CHALLENGE could mean giving up buying clothes and grooming products. Or it could mean saying "no" to snacks and takeout drinks.

Whatever you stop buying, the challenge will give you the chance to change your shopping habits.

You'll create less waste AND save money, too!

REUSING AND RECYCLING

5

One of the best ways to cut down on waste is to **reuse** and **recycle** whenever you can. This chapter will give you ideas and guidelines for:

> → Taking care of things you own.
>
> → Finding new uses for things you're not using.
>
> → Recycling things you no longer need.

TAKING CARE

In the past, most people took great care of their possessions. They polished and cleaned and mended to make sure things lasted as long as possible. Today, we can learn a lot from this approach. Instead of chucking things away, it's worth putting in the effort to look after your possessions.

I polish my boots to keep them looking good.

I have a wooden box that I've mended and varnished myself. I use it for my jewelry.

I've sewn some patches onto my jacket. I think they make it look cool.

I go to our local bicycle repair café. Now my bike's in great shape _and_ I've learned new skills.

REPAIRING SKILLS

If you want your things to last, sooner or later they'll probably need some repairs. Of course, some jobs need a professional, but there are many repair skills you can learn yourself.

Maybe you have friends or relatives who would love to share their knowledge with you? Or perhaps you can find some repair and maintenance classes?

Once you've learned the skills you need, you can give your possessions a new lease on life. You'll save a pile of money, and you can feel happy that you're helping the planet.

I wear my patches with pride. They show how much I care about the Earth.

GIVING THINGS NEW LIFE

Just because you've had enough of something, it doesn't automatically make it useless. Next time you go to throw something away, try asking yourself:

How can I find a new use for this?

With just a little creativity, you can think up all sorts of ways to reinvent your stuff...

Adapting something for reuse is sometimes known as **repurposing** or **upcycling**. Here are just a few examples:

OLD BUCKET   ⟶   PLANT POT

BROKEN PLATES ⟶ MOSAIC

PLASTIC BOTTLE ⟶ BIRD FEEDER

UPCYCLING IDEAS

You could keep a stash of things to upcycle —
cards, buttons, foil wrappers and all kinds of
other stuff you'd normally throw away. Then
you can have fun making them into attractive
gifts. Go to Usborne Quicklinks (see page 9) for
links to websites with ideas and instructions.

TIME TO RECYCLE!

Once you're sure you can't use something anymore, it's time to **recycle** it, if you possibly can.

Did you know that almost <u>two thirds</u> of everything we throw away could be recycled? Each time you go to chuck something away, remember to ask yourself:

Could this be recycled instead?

WHY RECYCLE?

Recycling is <u>essential</u> for our future.
Not only does it help cut down
on waste, it also has many
advantages for the health
of the planet.

➜ Recycling saves energy.

Making products from recycled materials consumes
less energy than making the same things from
raw materials. It also avoids using energy to
extract and process new materials.

➜ Recycling reduces landfills.

Each time we recycle, we're helping to tackle
the problem of waste disposal, and reducing
the threat of pollution from landfills.

→ **Recycling helps protect the environment.**
When we recycle used materials (such as paper, glass, metal and plastic) we're reducing the need for mining, quarrying and logging – activities that damage natural habitats. And by cutting down on processing raw materials, we're also producing fewer carbon emissions.

GETTING ORGANIZED

Recycling programs are usually run by local governments, and they can vary across the country. Some cities give out leaflets explaining what they collect and when, and you can always check out your local program online.

SORTING AND SEPARATING

In some places, you can lump all your recycling together in one container. In others, you'll be asked to separate metals, glass, plastics and paper. It is <u>very important</u> to sort things if you're asked to. Otherwise, they may end up being buried or burned instead of being recycled. To make sorting easier, you can use different bins for different kinds of recycling.

SO, WHAT CAN YOU PUT IN YOUR RECYCLING?*

Take a look at the next few pages to see what you **CAN and CAN'T** put in your recycling bins...

Some things might surprise you!

* Guidelines can vary from place to place, so it's REALLY IMPORTANT to check your local recycling website.

YES

✓ Newspapers and magazines

✓ White and colored paper and cardboard

✓ Brochures, catalogs and junk mail

✓ Envelopes (including see-through windows)

✓ Greetings cards (NOT glittery cards though)

✓ Cardboard tubes from toilet paper and paper towels

✓ Wrapping paper (NOT glittery or metallic)

✓ Cardboard food and drink cartons (thoroughly rinsed)

NO

X Used paper towels or tissues (to avoid spreading germs)

X Glued or painted paper

YES

✓ Yogurt, margarine and ice cream containers

✓ Cleaning product bottles

✓ Drink bottles, milk cartons, bottle tops

✓ Shampoo bottles

✓ Food cartons, trays and packaging (but NOT polyethylene film)

Make sure all plastic items are washed thoroughly.

NO

X Hard plastic items, such as storage boxes

X Plastic toys or gadgets

X Polystyrene foam

X Chips bags

GLASS

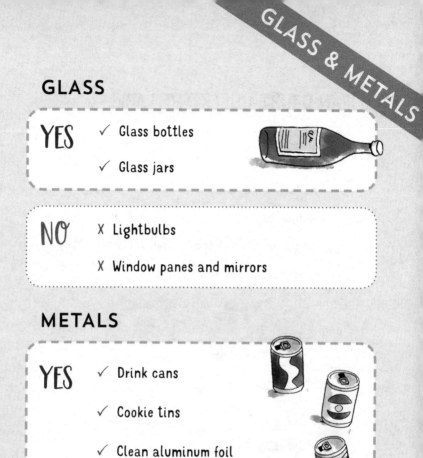

YES   ✓ Glass bottles

      ✓ Glass jars

NO   X Lightbulbs

      X Window panes and mirrors

METALS

YES   ✓ Drink cans

      ✓ Cookie tins

      ✓ Clean aluminum foil

      ✓ Metal food cans

      ✓ Metal lids and bottle tops

NO   X Cooking pots or heavy metal objects
      (Take to a recycling center – see page 92.)

IF IN DOUBT, <u>CHECK THE LABEL</u>

The two symbols shown below are used around the world to indicate that an item can be recycled (although not all recyclable items have a label).

 This symbol shows that an item CAN'T be recycled.

 This symbol tells you to check the guidelines on your local recycling website.

Some labels show you what to do with your trash, such as "Rinse before recycling."

 This explains that after rinsing you should put the can's lid inside the can.

This describes the object to be recycled.

Some labels show that an item is made up of different parts that need to be treated in different ways.

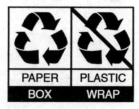

This sign (actually two-tone green) is sometimes confused with the green recyclable symbol. It shows that the producer has paid money towards a recycling program, but it DOESN'T mean the item can be recycled.

RECYCLING ALERT!

! Make sure your recycled items are rinsed clean and free from food.

If your recycling smells bad, it will be rejected.

! Never put your recycling in a black bag. If the collectors can't see what's inside, your bag will automatically go to a landfill site.

! Keep recycling bin lids firmly shut so animals don't get injured.

! Check that your recycling DOESN'T include any pet litter, medicines, sanitary products or diapers. If any of these items are found, they can cause an <u>entire truckload</u> of recycling to be sent to a landfill site.

WHAT ABOUT FOOD WASTE?

A growing number of recycling programs will collect food waste from your home. Some food waste is compressed at a very high heat to produce **compost** for use in farming. Some is converted into **biogas** by a process known as anaerobic digestion.

The food waste is collected and stored in tanks, where it is broken down by microorganisms to release the biogas. Biogas can be used instead of fossil fuels to provide power for heating systems or vehicles.

WHAT CAN I PUT IN MY FOOD WASTE BIN?

YES

✓ Uneaten food and plate scrapings

✓ Fish (including bones and scales)

✓ Fruit and vegetables (including peelings and fruit pits)

✓ Meat and bones (cooked or uncooked)

✓ Dairy products (such as cheese)

✓ Teabags and coffee grounds

Remember to remove any plastic or clingfilm.

✓ Eggs and eggshells

✓ Bread and pastries

NO

X All non-food products

X All food packaging

X Liquids such as milk (These may leak when the waste is being transported.)

FOOD WASTE OR COMPOST?

Even if you make your own compost, you'll still need

to have some food waste collected. **Meat,**
fish, cheese and other **dairy products**
shouldn't be added to compost because
they attract rats, so you'll need
to put them in your
food waste bin.

Mmmm ...
smells like
dinner!

YARD WASTE

Some recycling programs collect yard waste and turn
it into compost. Use your yard waste bins for **grass**
cuttings and **hedge trimmings, leaves, plants,**
weeds and **small shrubs.** You can also put
compostable bags into your yard waste,
but the bins shouldn't
be used for soil,
bricks and rubble
or large branches.

WHAT ELSE CAN I RECYCLE?

Some things need to be taken to special collection sites for recycling. Check on the internet to find recycling centers near you and find out how to recycle household items such as:

WOOD AND FURNITURE   Bulky yard waste

Cooking pots and other heavy metal objects

ELECTRICAL APPLIANCES (big and small)

Engine oil – in a sealed container

Recycling banks can sometimes be found in supermarket parking lots or entrances. They often have bins for:

CLOTHES   SHOES

Bed linen

TOP-PRIORITY RECYCLING

Batteries, phones and **printer cartridges** all contain
highly toxic substances that can be dangerous if they
end up in landfills. But the good news is that all
these tricky items can be recycled safely.

Batteries

Batteries can be recycled through your
local recycling center, and some home
improvement and electronics stores
have recycling programs or
drop-off kiosks.

Printer cartridges

Several organizations have programs to recycle
printer cartridges. They are often found in office
supply stores and offer credit toward new cartridges.

Phones

Many phone stores and other retailers have recycling points for phones. The phones may be reconditioned for use by someone else, or they may be recycled to extract reusable plastics and valuable metals. Phones contain small amounts of gold, silver, copper and platinum that can all be reused.

Recycling the metals in phones removes the need to mine for more metal and reduces the harm done to the planet.

NEW LIFE FOR OLD SPECS

You can contact your optician or charity organizations to see if they will accept old glasses. There are many non-profit groups that provide free glasses for people in developing countries.

GETTING DRASTIC ABOUT PLASTIC

6

Around a hundred years ago, scientists created an amazing new material called...

PLASTIC

It was flexible, lightweight, cheap to produce, and extremely long-lasting. People loved using plastic and manufacturers kept on making more.

Just think of all the things we can make!

It was very exciting, until people started to realize that plastic came with some SERIOUS drawbacks.

Um...maybe it wasn't such a good idea after all.

PLASTIC PROBLEMS

Plastic is one of the biggest causes of **pollution** on our planet, harming humans, animals and the environment. And the worst problem of all is ...

WE <u>CAN'T</u> GET
RID OF IT.

Basically, plastic **doesn't biodegrade**, which means it can't rot away naturally. Instead, it simply breaks up into smaller and smaller pieces. Plastic will stick around for thousands of years, and we keep creating more and more of it.

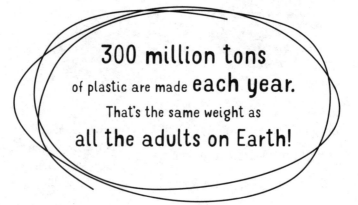

300 million tons
of plastic are made **each year.**
That's the same weight as
all the adults on Earth!

Plastic harms the environment when it's being made <u>and</u> when we try to get rid of it.

The process of making plastic **uses up valuable resources,** especially water.

Making plastic also **uses fossil fuels** and **causes air pollution.**

Because plastic doesn't biodegrade, it has to be buried or burned. This can release **harmful substances** into the environment and still doesn't get rid of it entirely.

Some plastics contain **toxins**, which can enter the bodies of people and animals through their noses, mouths or skin.

Plastic waste that's left lying around can **pollute water and food supplies** for animals and humans.

Clumps of **floating plastic waste** are creating giant "rafts" in our oceans. These areas are so toxic that nothing can survive in them.

WILDLIFE IN DANGER

All kinds of creatures are in danger from plastic waste. Even the tiniest fish can take in plastic particles suspended in water. Larger animals can get tangled up in plastic litter, or swallow plastic bags, drinking straws and other plastic objects.

Plastic bags are littered all over the planet and they are <u>especially deadly</u> to wildlife. If a bag gets stuck in an animal's gut, it can prevent the creature from absorbing food, and lead to a slow and painful death.

GOOD AND BAD PLASTIC

The best thing to do for the planet would be to stop using plastic altogether, but sadly it's not that simple. However would we manage, for example, without all our plastic medical equipment? On the other hand, there are countless plastic items we don't really need at all. Many of these items are "single-use," which means they're used just once and then thrown away.

Here are some examples of single-use plastics — the biggest cause of waste pollution on our planet.

drink bottles

plastic shopping bags

drinking straws

food packaging

takeout trays

LOOKING AT SINGLE-USE PLASTIC

Around <u>one third</u> of the plastic produced in a year is made into single-use items. And most of these items are NOT recycled.

Every minute of every day, an astonishing **one million** plastic bottles are bought around the world – and **less than half** of them are recycled.

It's clear that single-use plastic is a DISASTER for the planet and we need to take URGENT action.

If EVERYONE cuts down on single-use plastic:

→ We'll be making a big reduction in plastic production.

→ We'll be tackling the massive problem of plastic pollution.

WHAT'S BEING DONE ABOUT SINGLE-USE PLASTIC?

Recently, many people have been making determined efforts to cut down on single-use plastic.

COMPOSTABLE PACKAGING

Some manufacturers have switched to packaging made from compostable materials, such as potato starch.

PLASTIC BANS

Many governments have banned stores from selling or giving out plastic bags.

CHANGING STORES

Some stores provide compostable bags and a few are starting to sell loose fruit and vegetables without packaging.

ZERO-WASTE STORES

Zero-waste stores are on the rise. As well as selling loose fruit and veggies, they stock large quantities of dry and liquid goods and encourage customers to fill their own containers.

NO MORE PLASTIC STRAWS!

Plastic drinking straws are on the way out, and are being replaced by metal or paper ones.

RECYCLING ON THE RISE

The recycling of single-use plastic is on the rise in many countries, and a growing number of products are being made from recycled plastic.

This is all encouraging news, but we still need to CHANGE OUR ATTITUDE to plastic urgently. Luckily, there are some simple actions to take ...

TEN PRACTICAL STEPS TOWARDS ZERO PLASTIC

 1 **Take your own bags.**

Carry them with you everywhere so you never have to use a store's plastic bags.

You never know when a shopping bag will come in useful.

 2 **Choose soap bars.**

Go for old-fashioned bars rather than plastic containers of soap and shower gel — and look out for solid shampoo bars, too.

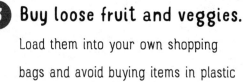

3 **Buy loose fruit and veggies.**
Load them into your own shopping
bags and avoid buying items in plastic
bags or trays.

4 **Find alternatives to plastic.**
Buy items made from wood, cardboard,
paper, glass or metal, instead of plastic.

5 **Shop the zero-waste way.**
Look out for zero-waste stores where
goods are sold loose or in bulk, and fill
up your own bags and containers.

6 Get your milk delivered in bottles.

Find out if there's a milk delivery service near you, and switch from buying plastic containers to having your milk delivered in glass bottles.

7 Avoid using plastic film.

Store items in the refrigerator in a bowl with a plate on top, or use beeswax wrap (cotton coated in wax) or biodegradable film.

8 Make a waste-busting kit.

Carry your own mug, water bottle, and food container so you can always say "NO" to plastic. Check out the waste-busting kit on page 70.

Reuse the plastic you've got.

Find new uses for your plastic, such as turning your old ice cream cartons into picnic boxes or freezer containers.

 As a final resort – recycle your plastic.

If you're sure you can't find any other use for them, recycle your plastic items. At least they'll be having a useful new life and <u>not</u> causing dangerous litter and pollution.

WATCH OUT FOR HIDDEN PLASTIC!

You can find tiny particles of plastic in some very surprising places...

These things all contain **plastic microparticles**:

wet wipes

glitter

TEABAGS

CHEWING GUM

These all have **plastic linings**:

COFFEE CUPS

SOUP CARTONS

CHIPS BAGS

drink cans

And many synthetic or "man-made" fabrics are made from **plastic microfibers.** Here are a few examples:

nylon   ACRYLIC

POLYESTER   *LYCRA*

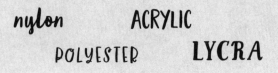

WHAT HAPPENS TO THE PLASTIC?

Each time you use a product containing plastic
particles, some of them are released into the air.
This happens especially when you wash or
dry clothes made from synthetic fabrics.
When the clothes are washed, plastic
microfibers enter our waste water. And
when they are put in a dryer,
fibers are released into the air.

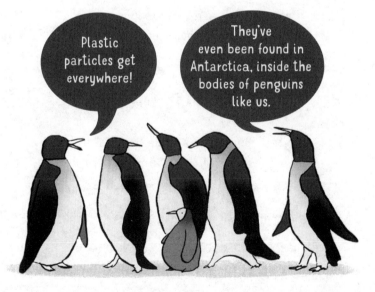

Plastic particles get everywhere!

They've even been found in Antarctica, inside the bodies of penguins like us.

FIVE MOVES TO BEAT "HIDDEN PLASTIC"

If you want to avoid the damage caused by hidden plastic, you'll need to start by changing your shopping habits.

- ✓ Stop buying wet wipes, glitter and glittery things.

- ✓ Don't buy chewing gum.

- ✓ Buy plastic-free teabags or go for loose tea.

- ✓ Try to avoid drink cans, takeout coffee cups and cartons.

- ✓ Buy clothes made from natural fibers, such as bamboo or wool.

All these actions will help make a difference, but what do you do with your clothes made from synthetic fabrics?

CAREFULLY DOES IT

It's clearly not a great idea to throw away your clothes. (You can check out the drawbacks of throwing away clothes on page 65.) Instead, the best approach is to wash and dry them with care — especially as older clothes release fewer particles than new ones.

You can reduce the number of particles being released if you wash your clothes at a low temperature and dry them naturally (rather than in a dryer). You can also buy special clothes bags that act as micro-filters, keeping most of the plastic particles out of the waste water.

I'm doing my best to limit plastic damage.

WHAT ELSE CAN WE DO TO BEAT THE PLASTIC PROBLEM?

So, you're cutting down on plastic in your daily life, but what can you do to make a wider impact? Here are a few plastic-busting actions:

I've organized a team to clean up all the plastic from our local streets.

I raise money for environmental groups that tackle plastic pollution worldwide.

I write to politicians urging them to get drastic about plastic.

I campaign to ban plastic packaging in stores.

SHOPPING CHOICES

7

Each time we go shopping, we ask ourselves some questions to help us make good choices:

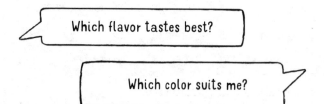

Which flavor tastes best?

Which color suits me?

And there are other questions we can ask as well:

Was this produced in a way that harms our planet?

Is this made of materials that can be recycled?

Once you start asking questions like these, you can use your shopping choices to help take <u>better care</u> of our planet.

SHOPPING POWER

Your choices might not seem important on their own, but when they're put together with other shoppers' choices, they can be part of a movement towards more planet-friendly shopping.

Retailers and manufacturers react to what's being bought and what's being left on the shelves, and adjust their products to sell what their customers want.

How will YOU use your super shopping powers?

This means <u>we can all make a difference</u> to the products being made, and to the things being sold in our stores.

FOUR SIMPLE SHOPPING GUIDELINES

1 Choose eco-friendly products.

"Eco-friendly" stands for "ecologically friendly." It means that the products are made with as little harm as possible to the Earth.

★ Eco-friendly cosmetics, detergents and other products are made from substances that <u>don't</u> contain harmful chemicals.

★ Some eco-friendly products are made from plants or trees, such as bamboo or pine, that are very fast-growing so they're not in danger of running out. Bamboo products include clothing, sports gear, toothbrushes, and make-up remover pads. (These pads can be washed and reused many times.)

> Eco-friendly products are sometimes known as **eco products** or **sustainable products**.

Look out for things made from recycled materials.

You'll find a growing range of recycled products for sale. Here are just a few to look out for:

★ Toilet paper, paper towels and cards made from recycled paper

★ Drinking glasses made from recycled bottles

★ Jewelry and baskets made from recycled magazines

★ Furniture made from recycled wood

SPOT THE SYMBOL
This symbol shows the percentage of recycled materials in a product.

Buy kinder, fairer products.

Caring for our planet means caring for the people and animals living on Earth.

 Look out for products with the Fairtrade label. This shows the people involved in making or growing the products will get a fair share of what you've paid.

 When you're buying meat, eggs or fish, **think about how the animals have been treated.** You can read about food production methods on pages 125 to 131.

> There's much more on food shopping in Chapter 8.

☑ **Check for the "leaping bunny" logo.** This guarantees that no new animal tests have been used in the development of a product.

4 Buy or make planet-friendly gifts.

Why not make your gifts a way of caring for the Earth? Look out for items made from sustainable materials and buy handmade objects crafted by local makers rather than factory goods that may have been transported very long distances. That way you'll be reducing the harm done to the Earth.

MAKING, NOT SHOPPING

If you make your own gifts, instead of going shopping, you'll be cutting down on energy-use and pollution. Go to Usborne Quicklinks (see page 9) for links to websites that show you how to make some simple handmade gifts.

I've made a memory box.

I've made a recipe scrapbook.

I've knitted a scarf.

I've baked a birthday cake.

I've made a bookcase.

118

AND AVOID PLASTIC TOO ...

As a planet-friendly shopper, you should always try to avoid:

 Anything made from plastic

 All unnecessary packaging – <u>especially</u> when it's made from plastic

PACKAGING PROTEST

Some people choose to make their feelings about packaging very clear. After they've bought their items, they remove the packaging and leave it at the checkout before packing their own bags. Another approach is to buy unpackaged products elsewhere ...

The protesters hope that stores will react by putting pressure on manufacturers to reduce their packaging.

ONLINE SHOPPING

Of course, the most planet-friendly way to shop
is to walk or cycle to your local stores, but you'll
probably still want to buy some things online.

Shopping online cuts down on car trips to the stores,
but delivering orders to your home involves road
transportation too. It's good to be aware of the costs
to our planet before you press the "buy now" button.

THREE GREENER WAYS TO SHOP ONLINE

> If only more people used pick-up points, I could cut way down on my individual drops.

1 Instead of having your goods delivered to your door, <u>collect your items in store or at a pick-up point</u>. Stores have regular deliveries, so adding your package won't cause much more damage to the environment — especially if you walk or cycle to collect it.

Hmmm. Is it vital to
have this backpack
tomorrow?

2 Ask yourself, "Do I really need
next-day delivery?" Trucks with
express deliveries usually have
to travel to places much further
apart than they do for standard
deliveries, and this all adds
to the traffic and pollution
on our roads.

3 <u>Try to avoid returning things.</u>
Think carefully before you place
your order and follow online
sizing guides for clothes, rather
than ordering several sizes and
sending some of them back.

It has been estimated that **one in five**
of all clothing items ordered online are
returned. That's an awful lot of journeys
to add to the damage to our planet.

8

FOOD CHOICES

The way our food is produced and the things we choose to eat have a big impact on the health of the Earth. If you want to help your planet, you need to make choices that are:

- Good for the Earth
- Good for the people and animals living on Earth
- And good for <u>you</u>.

To help you make these choices, it's important to have some knowledge of how your food is produced and how it reaches your plate...

Hmmm, I've never stopped to think about where my dinner comes from.

"INDUSTRIAL" FARMING

In today's world, farming can be BIG business as farmers compete to produce huge amounts of food efficiently and cheaply. Some farmers run very large-scale farms and aim to get the maximum amount of food from their land. To achieve this goal, they often use methods that damage the environment, are unkind to animals and harm human health. This type of agriculture is sometimes known as "industrial" farming.

ADDING CHEMICALS

Most industrial farmers spray their crops with pesticides and herbicides to kill insects and weeds, and add fertilizers to the soil to encourage growth. These artifical chemicals can damage wildlife and pollute groundwater, and the traces that remain on the food may be harmful to humans.

ANIMAL ISSUES

Some farmers rear animals on a very large scale. In order to make big profits, they feed their animals low-quality food and prevent them from roaming freely. Animals kept in cramped conditions are at risk of becoming ill, so they are given daily doses of antibiotics with their food and water.

Farms where animals are raised in crowded spaces are often known as **factory farms** or **battery farms**. This kind of farming results in poor animal health, low-quality meat, and threats to human health.
AND WE DON'T LIKE IT!

Of course, not all farmers use "industrial" methods. Many take a more responsible approach to their land and animals, and some choose to farm the "organic" way.

I look after the planet and grow tasty food.

ORGANIC FARMING

Organic farmers use methods that are as **natural and non-polluting** as possible. They use lower levels of pesticides than non-organic farmers and avoid all artificial fertilizers and herbicides. Organic farmers also aim to protect the natural environment and allow wildlife to flourish.

Animals on organic farms enjoy far more freedom and consume more wholesome food than those on industrial farms. This makes for much healthier animals, and healthier food for humans, too.

We love the outdoor life!

FREE RANGE = ORGANIC?

Meat, eggs and dairy products from organic farms are also labeled "free range" because the animals have freedom to roam. But the term "free range" <u>doesn't</u> automatically mean organic. For example, free-range eggs may come from hens that are given non-organic feed to eat and are kept in large cages with only limited freedom to move around.

I'm free-range AND organic!

SEA CREATURES IN TROUBLE

In recent years, the world's fish stocks have been in steep decline. Overfishing has meant that fish are caught before they have the chance to mate and reproduce. And modern fishing techniques can be very wasteful. Many fish are caught by mistake and thrown back dead into the water.

Dolphins and turtles can get tangled up in fishing nets and drown. And all sea creatures are at risk from plastic waste and from other forms of water pollution.

TAKING MORE CARE

Fortunately, there have been some positive moves towards responsible fishing and greater care for wildlife in our rivers and seas. Here are some examples:

The Marine Stewardship Council (MSC) has established clear **guidelines for sustainable fishing.** These guidelines state that fish must be caught in water with adequate fish stocks and that other creatures mustn't be harmed in the process.

Most tuna fishing boats have adopted methods that limit harm to dolphins and other large sea creatures. They avoid the parts of the ocean where dolphins usually swim, and use a "pole and line" fishing method designed to catch just one tuna at a time.

In some parts of the world, conservation teams are leading clean-up projects and tackling pollution in rivers, lakes and seas.

PROCESSING FOOD

Once food has left the farm or the fishing boat, it is usually taken to some kind of processing plant. There, some foods are simply washed and packaged, while others are combined to make "processed" foods, such as pies, pizzas or cookies. Processing food uses up huge amounts of energy, and sends polluting fumes into our atmosphere. And the food is often packed in plastic, adding to the problems of our planet.

FOOD ON THE MOVE

After food has been processed and packaged, most of it still has a long way to travel to reach the stores. Every day, food travels across the world in planes, trains and trucks, using up energy and creating the carbon fumes that help to cause climate change.

THINKING ABOUT "FOOD MILES"

The term "food miles" is used to compare the journeys of different food items from the place where they were farmed or fished to the place where they are eaten.

For example, an apple grown on the other side of the world may travel

12,000 MILES

to reach your plate, while a locally grown apple may cover

FEWER THAN 5 MILES,

using much less energy and causing less pollution.

Freshly picked by me this morning!

The number of food miles traveled by an item of food is just one of the factors we need to think about when making choices about what to buy. You can check food labels to see where things come from.

A MEATY PROBLEM

One of the biggest food issues facing us today is the problem of MEAT. People around the world are consuming more and more meat, and meat production keeps rising to keep up with demand. This makes climate experts very worried. They have reached the conclusion that large-scale meat production is causing major damage to our environment and we need to tackle the problem <u>urgently</u>.

SO, WHAT'S THE MATTER WITH MEAT?

Rearing animals for meat — and growing the food to feed them — is a major cause of greenhouse gas emissions on our planet.

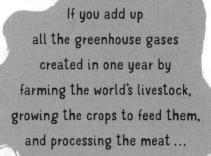

If you add up all the greenhouse gases created in one year by farming the world's livestock, growing the crops to feed them, and processing the meat ...

...they will equal the annual emissions from **all** the world's planes, trains, trucks and cars.

BURP ALERT!

Not only does animal farming use up fuel and create pollution — the animals themselves are adding to the problem. As cows and sheep digest their food, they burp out methane, a gas that contributes to climate change.

Pigs also produce methane, although in smaller quantities than cows and sheep. Still — it all adds up!

USING UP LAND

Animal farming doesn't just cause pollution — it also uses up VAST areas of land for growing animal food. If we want to feed all the people of the world, much of this land would be far better used to grow food for humans, rather than animal feed.

LOSING RAINFORESTS

The problem of using up land is especially serious in rainforest areas, where thousands of acres of forest are destroyed each year to create cattle ranches and fields to grow animal feed. Rainforests play an essential role in soaking up carbon dioxide from the atmosphere, so losing them has a SERIOUS impact on the health of our planet.

You can read more about our shrinking rainforests in chapter 11.

PALM OIL PROBLEMS

Rainforests aren't just under threat from cattle ranchers. They're also being cleared to make way for huge palm oil plantations. Palm oil is found in a wide range of processed food as well as in household items such as laundry detergent, toothpaste, soap and shampoo.

CHOCOLATE
ICE CREAM
COOKIES
INSTANT NOODLES
PIZZA

These are just a few of the processed foods that may contain palm oil.

Growing palm oil harms the environment, as forests are cleared and wildlife is destroyed. But efforts have been made to improve farming practices. Many palm growers have joined the Round Table on Sustainable Palm Oil (RSPO) (see page 145). It's not the perfect solution, but RSPO growers promise not to destroy important wildlife habitats, to treat their workers fairly, and to limit damage caused to the environment.

FOUR PLANET-FRIENDLY EATING CHOICES

If we <u>all</u> make some changes to our eating habits, we can REALLY make a difference to the health of our planet. Here are a few simple guidelines to follow:

1 First of all...

eat less meat

Aim for a mixed diet, and cut down on meat — especially beef, lamb and pork (often known as red meat). It may take a while to change your eating habits, but you can enjoy trying out different foods.

We have meat-free Mondays.

I just have meat as a treat. That way I enjoy it much more.

I've cut out meat completely, but I get enough protein from other foods.

2 Choose fish or poultry rather than beef, pork or lamb.

These foods can be fished and reared with much

less harm to the Earth than red meat.

They provide healthy,

low-fat ingredients

in a balanced diet.

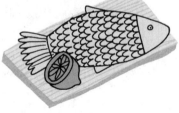

Eat more vegetables.

Growing vegetables creates
fewer carbon emissions than
rearing animals – and everyone
knows that veggies are
GOOD for you!

We contain
many essential
vitamins and
minerals.

 **Start eating more nuts, beans,
lentils and other protein-rich,
plant-based foods.**

Lentils, nuts, beans and peas all provide valuable
sources of protein. Growing them is an efficient
use of land, and has much less impact on the
environment than raising animals.

PROTEIN SOURCES

People often ask, how do you get enough protein if you give up eating meat? Meat is a useful source of protein, but there are plenty of other ways to get all the protein you need to keep you well. All the foods below contain high quantities of protein:

cheese

peanut butter

nuts

milk

tofu

BEANS

FISH

eggs

peas

LENTILS

Who'd have thought something so small could be so good for you!

SIX FOOD-SHOPPING GUIDELINES

You may not get involved in many food shopping choices, but here are some simple guidelines to follow. Try sharing these ideas with your family — you could end up persuading them to change their shopping habits!

 Aim to buy food that's local and "in season."

Fruit and vegetables that have been grown naturally in your local area <u>won't</u> have been forced to ripen under powerful lights and <u>won't</u> have had to travel for long distances. This means they will have used up less energy and created less pollution. They'll also taste fresher and have more flavor!

2 Grow your own fruit and vegetables.

Food you have grown yourself will probably be the freshest you'll ever eat. Read about growing your own on page 189.

3 Choose sustainable fish.

Check to see if fish and seafood have been fished sustainably. Look out for the blue Marine Stewardship Council sign and go for tuna caught by "pole and line."

4 Try to avoid processed food and food in plastic packaging.

Food processing uses up energy and creates pollution, while plastic packaging causes serious harm to the environment.

5 **Look for sustainable palm oil.**

The spread of palm oil plantations has led to the destruction of huge areas of rainforest, but sadly palm oil is very hard to avoid completely. However, you <u>can</u> look out for RSPO certified oil that's grown with reduced harm to the environment (see page 138). (Around 20 percent of all palm oils produced globally are RSPO certified.)

RSPO stands for <u>R</u>ound Table on <u>S</u>ustainable <u>P</u>alm <u>O</u>il.

Look for its palm tree logo on food labels.

6 **Consider buying organically farmed produce.**

Organic food is farmed with less harm to the Earth and to animals than food from non-organic farms. It's <u>not</u> loaded with artificial chemicals — and it tastes great!

WHAT ABOUT THE COST?

Food that's good for the environment can cost more, but many people think it's tastier and worth it. And you can reduce your costs by making a BIG effort to cut down on food waste.

WASTING FOOD

Have you ever thought about all the food we throw away? Even if we manage to eat every scrap on our plates, there's still the stuff that some of us clear out from the refrigerator because we haven't managed to eat it up before it goes bad.

Must get one for the trash!

WHAT A WASTE!

The average US family wastes at least **one third** of all the food they buy.

This is a SHOCKING WASTE in a world where millions are going hungry. And wasting food means wasting energy too. Whenever you throw away food, you're also wasting the energy that has been used to produce it and transport it to your home.

FOUR WAYS TO FIGHT FOOD WASTE

What can we do to make the most of our food and cut down on waste? Here are some guidelines for you and your family:

 1 **Decide on a meal plan before you shop.**

If you spend a little time planning ahead, you won't end up buying heaps of food you don't need.

Monday - chili
Tuesday - pasta
Wednesday - fish
Thursday - risotto
Friday - stir-fry

Go to Usborne Quicklinks (see page 9) for links to some delicious waste-busting recipes.

MORE THOUGHTS ON FOOD AND COOKING

So, you've made your planet-friendly choices, and you're doing your best to stop wasting food. But have you thought about how you and your family cook and store your food?

Turn the page to see some easy ways to stop wasting energy in the kitchen ...

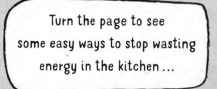

FIVE ENERGY-SAVING TIPS IN THE KITCHEN

1 **Cool down and warm up food naturally.**

Leave hot food to cool <u>before</u> you put it in the refrigerator or freezer. And remember to take out frozen food well in advance. That way it can defrost naturally, instead of being blasted in a microwave.

2 **Use a steamer for vegetables.**

Steaming uses less water than boiling, and it makes your veggies taste fresher too. If you're boiling potatoes, try steaming your veggies on top of them to make the most of your cooking energy.

Shut that door!

Resist the temptation to keep opening the oven door. Each time you take a peek, your oven will need an energy boost to regain its heat.

Only boil what you need.

When you're cooking on the stove, don't boil more water than you need. Choose a pan that's not too big, use just enough water to cover the food, and don't forget to use a lid to keep in the heat.

Cook large quantities.

Make the most of the heat by cooking large quantities of food. Then save portions to freeze or eat the next day.

9. ▶ GETTING AROUND

A hundred years ago, many people didn't ever travel far from home. But fast, efficient transportation has changed the way we live. In most parts of the world, it's normal to make daily trips by car, bus or train, and many people travel by plane several times a year.

It can be all too easy to jump into a car or head off on a plane for a vacation. But if we really care about the future of our planet, we need to take a careful look at how we get around.

What is all this travel doing to the planet?

TRAVEL TROUBLE

Planes, trains and ships and most types of cars consume fossil fuels and create the carbon emissions that lead to climate change. This is a growing problem, as more planes take to the skies each year, and our roads keep filling up with vehicles.

Road traffic also poses a threat to health, as vehicles pump out a mixture of harmful gases. In areas with heavy traffic, fumes from gasoline-fueled vehicles can cause breathing difficulties, and diesel fumes are especially harmful.

Transportation is the world's fastest-growing source of carbon emissions and general air pollution.

FINDING TRAVEL SOLUTIONS

Today, many people are waking up to the problems caused by our ways of getting around. There are efforts to reduce the amount of transportation we use, and to find cleaner and greener ways to travel.

➜ Scientists and engineers are developing a range of vehicles that don't use gasoline or diesel, with electric cars leading the field.

➜ Some governments have passed laws to phase out vehicles that run on diesel fuel. There are also plans to stop making gasoline cars within the next 20 years.

→ Some governments are rethinking the design of cities and towns, with more space given to bus lanes, walkways and cycle tracks.

→ Roadside charging-points are being installed, as people start to switch to electric vehicles.

→ Low-carbon trains, trams and buses are being introduced in many towns and cities.

→ Individuals, families and businesses are making deliberate efforts to cut back on unnecessary journeys.

We all make journeys in our daily lives, so we ALL need to think hard about our travel choices.

FOUR GREEN
TRAVEL CHOICES

 Walk whenever you can.

Walking is good for your health and for the planet – and it can be an excellent way to spend more time with your friends.

Maybe you (and your adults) could organize a **"walking bus"** for your journey to and from school? It's a healthy, safe and sociable way to travel.

You could talk to your teachers about having a monthly **"walk to school day"** when as many as possible make the journey on foot.

2 Get your wheels out.

Show you care for your planet by getting around with your friends by bike, scooter or skateboard. And take good care of yourself as well, by always wearing a helmet.

SKATERS FOR THE PLANET!

3 Use public transportation.

Taking the bus or train with other passengers is much more energy-efficient than making the same journey in lots of separate cars.

4 Organize a shared school run.

If you <u>have</u> to travel by car, make sure it's filled with friends to make the most of all the energy it's using.

WHAT ABOUT VACATIONS?

When it comes to vacation plans, you could try suggesting some planet-friendly choices ...

Why don't we explore somewhere closer to home, instead of jetting off by plane?

Bus and train journeys can be a lot more interesting than plane trips.

Let's put our bikes on a bus and have a cycling vacation!

Trains can be a great way to explore new places. On average, they use just one **third** of the energy consumed by a plane making the same trip, and cause about one **fifth** of the pollution.

10. LOOKING AFTER THE EARTH'S WATER

You may be one of the lucky ones who can turn on a faucet any time you want, and get all the clean water you need. But, for many people, life is not so simple. In some parts of the world, children have to trek for miles every day to fetch water for their families — and their water sources are often polluted.

One in three people on our planet don't have enough water for their needs, and **one in ten** can't rely on a clean water supply.

NO FAIR SHARES

Thousands of people in the world exist on **less than 3 gallons** of water **a day.**

Meanwhile, the average person in the US uses **80 to 100 gallons a day.** (Makes you think, doesn't it?)

A GROWING PROBLEM

As our climate changes, the water situation is getting worse. Many places have much less rainfall than they used to. Without enough rainfall, harvests fail, and the lack of clean water leads to widespread disease. In other places, there are serious floods, which also threaten farming and human health.

WASTING WATER = WASTING ENERGY

In many parts of the world, water needs to be used with the greatest of care. But even in countries with plenty of water, wasting water still means wasting energy. All the water that ends up in our drains has to be decontaminated (cleaned) before it's used again. This process uses LOTS of energy and so creates damaging carbon emissions.

DOWN THE DRAIN!

When you think how precious water is, it's shocking we don't treat it with more care. All too often, we let it go down the drain without being used at all!

TACKLING WATER SHORTAGES

Faced with the problems of water shortages, many people are working on ways to conserve it.

Scientists are developing new ways to capture and store water. Water companies are working to prevent water loss from leaking pipes, and water charities are trying to maintain clean water supplies in countries suffering from drought. Meanwhile, a growing number of farms and factories have introduced water-saving methods and machinery.

These are all positive moves, but it's up to ALL OF US to play our part in tackling water shortages.

We can ALL make an effort to use less water.

SEVEN WATER-SAVING MOVES

1 **Turn off the faucet.**
Get into the habit of turning off the faucet while you're brushing your teeth. Each time you leave the water running while you brush, you're wasting around a gallon and a half of water.

2 **Take a short shower rather than a bath.** And turn off the water while you soap yourself.

I sing one song as I shower – and I'm out before it's done.

3 **Use rainwater to water your plants.** You can set up a rain barrel or a row of buckets to capture the rain.

Did you know that a garden hose or sprinkler can use almost as much water in **one hour** as a **family of four uses in a whole day?**

Treat your indoor plants to an occasional rain bath, and when you're choosing plants for your house or garden look out for ones that <u>don't</u> need much watering.

Only wash FULL loads of laundry or dishes. Half loads are a waste of water, energy and detergent.

 Don't wash veggies or rinse dishes under a running faucet. Use a bowl instead and reuse the water on your plants.

I've saved this water especially for you.

6 Think about flushing.

Did you know that over a quarter of all the water consumed in your home is used to flush the toilet?

Some families choose to save water by only flushing after they poop. Cutting down on the number of times you flush the toilet is a very good way to save water, but it's something you'll need to discuss with all the family.

7 Talk to your family about water-saving moves.

Check that your faucets and showers aren't losing water through leaks and drips, and suggest changing to water-saving showerheads and toilets.

A leaky faucet that drips at the rate of **one drop per second** can waste up to **2,600 gallons** of water a year...

...that's the same amount as an average person in the US uses in **a month!**

KEEPING OUR WATER CLEAN

Water shortages aren't the only water problems facing us today. There's also the MASSIVE challenge of keeping our water clean. The health of our water affects all living things because we all share the same water, which is constantly recycled in a natural process known as the **water cycle**.

IT'S ALL THE SAME WATER

In the water cycle, water falls as rain or snow and gradually travels underground until it finally ends up in our oceans. From there, the surface water evaporates, turning into clouds, before falling again as snow or rain. This means the same water is being recycled, over and over again.

At each stage in the cycle, water can be polluted by harmful substances in the air, water and soil...

So we need to fight pollution at every stage.

WATER POLLUTION

There are many threats to our planet's water, but these are the main causes of water pollution:

HUMAN SEWAGE

chemical waste from factories, farms and landfill sites

OIL SPILLS FROM SHIPS

HOUSEHOLD WASTE

LITTER

Dangerous substances can <u>enter directly</u> into our rivers, lakes and seas or they can <u>travel through the soil</u>, contaminating the water systems that run underground. These substances cause harm to plants, fish and other creatures, and once plants and wildlife have been affected, the food that humans eat is polluted too.

WE'RE RESPONSIBLE TOO

Each time we send anything down our drains — whether it's via a sink or a toilet — we are putting substances into our waste water which may end up in the natural environment. Some of the water passing through our drains leaks into the surrounding ground, and even though waste water is treated at sewage works, some substances still enter rivers, lakes and seas.

FIGHTING WATER POLLUTION

In recent years, many projects have been set up to tackle water pollution. Some are international, dealing with the problems of our oceans. Others are local, with volunteers cleaning up rivers, ponds and streams. But clean-up projects aren't enough on their own — we ALL need to be aware of our impact on our planet's water. You can find some simple ways to clean up your water act on the next few pages...

FIVE WATER CLEAN-UP GUIDELINES

1 **Go for planet-friendly washing and grooming products.**

Have you ever thought that your daily grooming routine could be sending harmful chemicals down the drain? Some shampoos, shower gels and face washes contain ingredients that can damage wildlife. Next time you go shopping, look out for "eco-friendly" products made from biodegradable or plant-based ingredients. You could also join the growing number of people who make their own planet-friendly toiletries.

MAKE YOUR OWN FACE PACK

You can make a soothing face pack using the natural ingredients of **oatmeal**, **honey** and **egg**.

- → Grind 4 tablespoons of rolled oats into a coarse powder, using a food processor or a pestle and mortar.
- → Add 1 tablespoon of honey and the yolk of an egg and stir the mixture to form a smooth paste.
- → Apply the paste to your skin and leave for 15 minutes.
- → Wash your face thoroughly and feel the difference!

Oatmeal has soothing ingredients to calm itchy skin.

Honey has antiseptic properties to help treat acne.

Egg yolks are rich in vitamin D, which softens dry skin.

(Don't forget to save the egg whites for cooking!)

2 Keep oil, fat and grease well away from the sink.

You can add small amounts of cooking oil to your food waste, but you'll need to collect larger quantities in a glass jar with a lid. Check online to see if there's a waste cooking oil collection service in your area. If not, put the jar (with its lid on firmly) into your general trash can.

3 Never pour medicines or pills down the drain.

Tablets, medicines and creams may contain dangerous substances that can leak into the ground and end up poisoning plants, animals and fish. Take any unwanted medicines to a pharmacist or drug take-back site where they will be disposed of safely.

I'm using planet-friendly products AND being careful not to waste water.

Choose gentle cleaning products.

Dish and laundry detergents, sink and toilet cleaners all add powerful chemicals to our water, and bleach is especially toxic to living things. Aim to avoid bleach entirely and choose eco-friendly cleaning products. Zero-waste stores stock these cleaning products in bulk, so you can simply refill your own container.

5 Think before you flush.

Non-biodegradable objects, such as wet-wipes, tampons and diapers, should NEVER be flushed down the toilet. They can combine with oil and grease to create "fatbergs" — massive lumps of sticky waste that block up sewage systems.

DON'T FEED THE FATBERGS!

Once a fatberg is wedged inside a sewage pipe, it takes a giant effort to get rid of it. Most of the work of breaking up a fatberg has to be done manually, by chipping away at it with power tools — a job that nobody should ever have to do.

Fatbergs the size of **six double-decker** buses have been found in sewers.

MAKING A BIGGER SPLASH

As well as doing your best to save water and fight pollution, you can work on **bigger** changes too. Here are just a few ways to help make a difference:

I'm taking part in a river clean-up project.

I'm putting pressure on factories to stop dumping waste in our water systems.

I run a water-saving campaign at school.

And we can all talk to friends and family about treating water with more respect!

11. CARING FOR THE NATURAL WORLD

What can YOU do to help the natural world? This chapter lists some ways to care for the Earth and its wildlife. But first it looks at the problems facing all life on our planet.

Climate change

It's clear that the Earth's climate is changing. Some parts of the world are becoming hotter and drier while others are getting more rain, and serious fires and floods have devastated large areas of land. Meanwhile, the polar icecaps are melting, causing sea levels across the globe to rise.

POLAR BEARS IN TROUBLE

Melting ice around the North Pole has meant that polar bears' hunting grounds are shrinking, and the number of surviving bears is falling fast.

Shrinking forests

We are losing trees at a frightening rate. Every two seconds, an area of forest the size of a soccer pitch is cut down, and one third of all the world's tropical rainforests have been destroyed within the last 50 years. Deforestation (losing trees) has the effect of speeding up climate change, because trees play a vital role in soaking up carbon dioxide from the atmosphere. And losing trees also leads to the loss of animals, insects and plants.

Scientists estimate that **over 100** rainforest species become extinct **every 24 hours**. Golden toads were declared extinct in 2004.

Widespread pollution

The way we humans live on the Earth has resulted in widespread pollution. **Artificial chemicals** used in farming and industry soak into the ground and enter rivers, lakes and seas. **Harmful gases** pollute the air, and **litter** threatens wildlife on land and in water. Evidence of pollution can be found on the highest mountains and in the depths of the oceans. It is especially noticeable in delicate habitats, such as coral reefs, which are slowly dying and losing their brilliant colors.

CORAL REEFS AT RISK

Coral reefs are home to more than a quarter of all sea creatures, but more than half the reefs are dying. Rising water temperatures, fishing, litter, water pollution and tourism are all responsible for damage to these precious habitats.

Declining species

We need a wide variety of wildlife to keep a healthy
balance of life on Earth, but now this **biodiversity**
is under threat. Across the globe, wildlife is at
risk from climate change and pollution. Human
activities, such as farming and building, have claimed
more and more land, leading to the destruction of
natural habitats. And hunting, fishing, and poaching
have also been responsible for the loss of species.

Today, over 6,000 mammals, birds, fish, amphibians,
reptiles, insects and plants are on the list
of **critically endangered species**.
This means they are facing a
very high risk of becoming
extinct. The list includes the
Sumatran tiger, which has
been hunted for its fur.

WILDLIFE UNDER THREAT

Here are just a few of the species whose time
on Earth seems to be running out:

AMUR LEOPARD   BLACK RHINO

HAWKSBILL TURTLE

MOUNTAIN GORILLA

DISAPPEARING BEES

The number of bees around the world is dropping
fast, mainly due to the use of artificial chemicals
in farming. This is a serious problem
because bees are needed —
along with butterflies, moths
and other insects — to pollinate
the Earth's plants and crops.

WHAT'S BEING DONE TO HELP THE NATURAL WORLD?

Fortunately, it's not all gloom and doom. Many people are working hard to preserve and protect our planet and its wildlife.

Across the world, people are campaigning to save endangered species.

The campaign to save the Giant Panda is a success story. In 2016, it was taken off the list of endangered species.

Many countries have launched campaigns to plant more trees.

Tree-planting campaigns include the replanting of large areas of tropical rainforest.

Governments have created protected areas where wildlife can thrive.

Today, wildlife reserves cover nearly 15 percent of the Earth's land.

Bans on hunting, poaching and overfishing have allowed some animals to make a comeback.

Thanks to campaigns to put a stop to whaling, whales can now swim safely in most parts of the world.

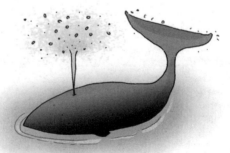

Teams of professionals and volunteers are working on clean-up projects.

Some clean-up projects have been very successful. Fish have returned to rivers that had become too polluted to support any life, and animals and birds are using breeding grounds that had once been abandoned.

Some polluting products have been banned.

To give just one example, it is now illegal on some Pacific islands to use sunscreens that damage coral reefs.

And people across the globe are taking steps to care for the natural world.

TEN WAYS TO HELP THE NATURAL WORLD

1 **Support conservation projects.**

Get together with friends to raise money
for your favorite cause. Some conservation
projects will even give you the chance to
adopt or sponsor an animal.

2 **Feed the birds.**

Set up a bird feeder or bird table and keep it
stocked with bird food. You could also provide
a birdbath filled with clean
water. Go to Usborne
Quicklinks (see page 9) for
guides to making simple
bird feeders and advice
on what to feed birds.

3 Grow some food.

When you grow your own fruit and veggies, you're producing delicious, healthy food, reducing the pollution caused by processing and packaging, <u>and</u> learning about the natural world.

GET GROWING!

With just a little knowledge and skill, you should be able to grow your own lettuce, beans and squash. Tomato and strawberry plants will thrive in large flowerpots, and you can create your own herb garden on a balcony or even a windowsill. Go to Usborne Quicklinks (see page 9) for some helpful growing hints.

4 **Organize a litter-picking and clean-up team.**

Cleaning up your neighborhood will give your local wildlife a better chance to thrive. Make sure you protect yourself by wearing gloves and brightly-colored clothing and use a litter-picker (a stick with a grabber that you operate by squeezing a handle). Some local government departments have litter-picking equipment that you can borrow free of charge.

5 **Enjoy the natural world.**

Try to spend more time enjoying the sights, sounds and smells of nature. The more you discover, the more you'll want to preserve our amazing natural world.

6 Help a local conservation group.

Many local groups welcome volunteers.
Check online to find a conservation project
near you.

7 Plant insect-friendly flowers.

You don't need a garden to plant
seeds — a window box or flowerpot
will work perfectly well. Go to
Usborne Quicklinks (see
page 9) for tips on
growing flowers that
attract butterflies,
bees and other insects.

8 Help plant trees and shrubs.

Look out for tree-planting programs to
support, and encourage your family and
school to get planting.

9 Start a compost bin.

Composting is an excellent way to improve the health of your soil <u>and</u> make good use of your family's food waste. Go to Usborne Quicklinks (see page 9) for links to websites on how to set up a compost bin and what to put in it.

10 Shop with care.

Before you hand over your money, think about the impact of the things you're buying on the natural world. If you're buying a paper product, for instance, look for a statement (like the one on the back of this book) that shows it has been made using paper from a sustainable source. You'll find a list of planet-friendly shopping tips on pages 115 to 119.

AND SOME ACTIONS TO AVOID...

→ **Never drop litter.**

A lot of street litter gets swept down drains and ends up in the sea. And even biodegradable waste can be bad for wildlife. Apple cores and banana skins, for example, can make some animals ill, so add them to your food waste or your compost bin.

→ **Don't pick wildflowers.**

Wildflowers are part of a natural habitat and should be left untouched.

→ **Don't buy seashells.**

Seashells may look pretty, but you should always remember they were once a creature's home. The trade in seashells encourages sellers to collect them, whether they're empty or not.

FIVE WILDLIFE SAFETY TIPS

Some types of trash can be a threat to animals if they're not recycled or disposed of carefully. Here are some ways to lessen the dangers, just in case some trash or recycling goes astray:

 Metal cans

Animals searching out food left at the bottom of cans can get their heads trapped or get injured by the cans' sharp edges. Remember to clean cans thoroughly before you recycle them.

 Plastic bags

Creatures can climb inside plastic bags and suffocate. You can help prevent animal deaths simply by tying a knot in the top of bags.

Plastic multi-pack drink can holders

Transparent drink holders are hard to spot so animals can easily get tangled up in them. Play it safe before recycling by cutting through all the plastic loops.

 Balloons

Brightly-colored balloons attract animals that may try to eat them and then choke to death. You can protect wildlife by cutting up balloons before throwing them away. Or simply don't buy them in the first place!

5 **Rubber bands**

Rubber bands can get wrapped around the beaks of birds or tangled around the bodies of small creatures. Keep them to reuse or, if you throw them away, cut them open first.

12. BUILDING A BETTER FUTURE

So — you've made lots of changes in your life, but what can you do to kick-start the BIGGER CHANGES that will help save our planet from further harm?

GET YOURSELF CLUED UP

Make sure you've done your research before you start talking to people. Check out the latest facts and figures, and stay up to date with recent news. With the right information at your fingertips, you'll be well prepared to drive home your points and deal with any arguments that might get thrown at you.

Once you know where to look on the internet, you can equip yourself with all the latest information. Go to Usborne Quicklinks (see page 9) for links to websites you can trust.

TALK THROUGH YOUR IDEAS

If you want to get things changed, a very good place to start is with your family and friends. Talking through your ideas with people you know will help you gain confidence before you start to reach out to the wider world.

PRACTICE YOUR PERSUASION SKILLS

The best way to get people on your side is to treat them with respect. Listen carefully to their point of view and try to respond to their concerns. Stick to the facts and avoid getting personal. And always try to keep calm, even when you're passionate about your beliefs.

FAMILY CONVERSATIONS

Try starting a conversation with your family about some planet-friendly choices. You could begin by discussing some simple changes, such as whether to switch to eco-friendly brands. Then you might move on to bigger decisions, like where to go for your family vacation.

Could we eat less meat?

Should we try walking and cycling more?

Can we save more electricity?

Should we grow our own vegetables?

GENTLY DOES IT

Don't be disappointed if your family isn't as excited as you are about radical changes. Try to take things gently, and remember it's very important to play your part if you want everyone to get involved.

Maybe you could volunteer to cook a meatless meal once a week, sort out the recycling, or start up a vegetable garden? This will show your family that you're really serious about making changes.

I'm in charge of our family compost bin.

GETTING TOGETHER WITH NEIGHBORS

Most people would love their neighborhood to be
a healthier and more attractive place. So why not
create a leaflet asking everyone on your street to
come to a meeting? You could suggest forming a
clean-up team to get rid of trash. And you could make
a plan to present to your local planners for planting
more flowers, shrubs and trees in your street.

WORKING WITH FRIENDS

It's good to share the effort when you're taking on
a new challenge. Try getting together with friends
to talk through the things that worry you — and the
changes you'd like to make. Maybe you could start
by sharing this book with them?

Taking action is so much easier when you're part of a
group. People will take more notice of the things you
do. You can share out the work between you, and you
can give each other support and encouragement.

Let's go to
the climate change
meeting together!

Yes — and let's
ask Josh and
Aisha too.

MAKING A DIFFERENCE AT SCHOOL

Think about forming a "green team" to make your school a more planet-friendly place. Put together a list of proposals and ask for a meeting with your school principal to talk through your ideas.

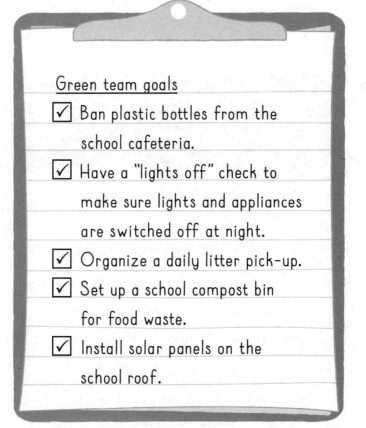

Green team goals
- ☑ Ban plastic bottles from the school cafeteria.
- ☑ Have a "lights off" check to make sure lights and appliances are switched off at night.
- ☑ Organize a daily litter pick-up.
- ☑ Set up a school compost bin for food waste.
- ☑ Install solar panels on the school roof.

CONTACTING YOUR CITY GOVERNMENT

Find the contact details of your city leaders and let them know your concerns about your area. Maybe you need some cycle lanes on your route to school? Or perhaps there's a local business that's creating polluting fumes? Send a polite email outlining the things that concern you.

You could get the email signed by friends and neighbors to show that your ideas have lots of support.

INVOLVING YOUR REPRESENTATIVES

If you want to change the way your state or country is run, your state representatives are the people to contact. They have a duty to speak out for the people they represent, so you need to make sure they hear your voice.

You can contact your city council by email or you can attend a community meeting. If you get the chance of a face-to-face meeting, make the most of the time by checking all the relevant facts in advance and noting down any questions you want to ask. Make your worries clear and urge your representative to vote for important measures such as replacing fossil fuels by greener energy sources and planting more trees.

PUTTING ON THE PRESSURE

You may decide some businesses need to be told that they're harming the planet. Maybe you could organize a **petition** to urge a company to cut down on packaging? Or perhaps you could ask some awkward **questions** about the ingredients a manufacturer uses? Sometimes, simply showing that a business is in danger of losing public support is all that's needed to prompt a change.

PETITION TO CUT DOWN ON PACKAGING

SHOWING THE WORLD

As the dangers facing our planet become clearer, more and more people are taking their message to the streets. In towns and cities across the world, there are demonstrations, strikes and marches, all of them urging action to save the Earth. And these protests are not just for adults — many of them are led by teenagers and children.

Our Home needs Help!

USING SOCIAL MEDIA

The internet is a powerful tool for spreading your message, but you'll need to use it carefully and not give out any personal details. You could set up an online group to share information with friends who have similar goals. Or you could create a blog, with comments on different topics and events in the news.

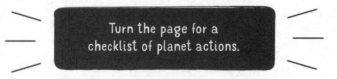

Turn the page for a
checklist of planet actions.

TEN STEPS TOWARDS PLANET ACTION

1. Do your research thoroughly – so you can act with confidence.

2. Work with friends – you'll be stronger together.

3. Talk through your ideas and plan your actions carefully.

4. Encourage your neighbors to create a cleaner and greener living space.

5. Work to make your school more planet-friendly.

6. Contact your city council about local issues.

7. Let your state representative know your views.

8. Put pressure on stores, manufacturers and businesses to clean up their act.

9. Use social media to spread your message.

10. Show the world you care by joining in marches and demonstrations.

AND REMEMBER...

Just because you're young, it doesn't mean you can't make a difference — or that you shouldn't be heard.

Sometimes older people need to be reminded about the things that really matter. After all, it's YOUR future that's being threatened, so it's only right that YOU should have a say in what's being done to protect it.

We need to WORK TOGETHER — young and old — to help our precious planet before it's too late!

IMAGINING THE FUTURE

What would life be like if we managed to make all the changes suggested in this book? As well as feeling secure about the future of the Earth, we could all look forward to living in a healthier, happier world.

The air we breathe would be fresher and cleaner.

Our food would be tastier and healthier.

THE COUNTRYSIDE WOULD BE FILLED WITH WILDLIFE.

Our cities would be more pleasant places to live.

WE WOULD GET MORE EXERCISE.

We would have fewer possessions, but we would value them more.

Of course, this is a very idealized picture, but it's still a goal worth aiming for. So, let's get working NOW to help save our wonderful planet!

PLANET WORDS AND TERMS

Here are some definitions of terms used in the book.

Words in **bold** have their own definition.

atmosphere

The mixture of gases around a planet.

biodiversity

The range of plant and animal life in any **habitat**.

biomass energy

Energy made from plant or animal waste. Biofuel and biogas are both examples of biomass energy.

carbon dioxide

The main gas produced by burning **fossil fuels**, and the main cause of the **greenhouse effect**.

carbon emissions

The gases released by the burning of **fossil fuels**. Carbon emissions are also known as "carbon dioxide emissions" and "CO_2 emissions."

carbon footprint

The amount of **carbon dioxide** released into the atmosphere as a result of the activities of a country, a business, a family or an individual.

clean energy

Energy that is produced without polluting the Earth's atmosphere. **Solar power** and wind power are examples of clean energy.

climate change

A change in climate patterns around the globe, and in particular a rise in the average temperature in many parts of the world. Climate change is sometimes described as global heating or global warming.

compost

Rotting plant material that can be used to add richness to soil.

compostable

Able to break down completely, leaving no pollution in soil or water. Compostable bags are often made from potato starch.

conservation

The protection and maintenance of an environment and its wildlife.

critically endangered

Extremely likely to become extinct (die out).

deforestation

Reducing or destroying forests by cutting down or burning trees.

ecological

Causing as little damage to the environment as possible, or even positively benefiting it. Ecological is often shortened to "eco" as in "eco-products" and "eco-settings" for appliances such as washing machines.

endangered

Likely to become extinct (die out).

fossil fuels

Fuels, such as gas, coal and oil, that were
formed underground from the fossilized
remains of plants and animals. When fossil
fuels are burned to produce power they
release **carbon emissions.**

freecycling

Freecycling is short for free recycling.
It involves giving away unwanted items
to people who can use them, instead of
throwing them away.

generator

A machine that converts the energy
of movement into electricity.

"green"

Causing as little damage to the environment
as possible, or even positively benefiting it.

green hydrogen

Hydrogen gas produced from **clean energy**
sources, such as **solar power** or wind power.
It is called "green" because it is produced
causing as little damage to the environment
as possible.

greenhouse effect

The trapping of the Sun's warmth in the Earth's lower **atmosphere** by a layer of gases. This results in raising the temperature of the Earth's surface, causing **climate change**.

greenhouse gases

The gases that create the **greenhouse effect**. They include **carbon dioxide** and **methane**.

groundwater

Water found underground in the cracks and spaces in soil, sand and rock. Groundwater eventually enters rivers, lakes and seas.

habitat

The natural home of an animal or plant.

incinerator

A furnace for burning waste material.

industrial farming

Industrial farming is sometimes called intensive farming or factory farming. It involves farming on a large scale with the help of artificial chemicals, and keeping large numbers of livestock in cramped conditions.

landfill site

A place where trash is dumped, and, in richer countries, buried under layers of earth.

low-carbon

Causing only a small release of **carbon dioxide** into the atmosphere.

methane

The gas produced by decomposing plant and animal matter or by burping cows and other animals. Methane is a **greenhouse gas** that helps to create the **greenhouse effect**.

organic farming

Farming that uses methods that are as natural and non-polluting as possible, and aims to support the natural environment and the health of livestock.

overfishing

Catching more fish than can be replaced naturally. Overfishing results in reduced numbers of fish and can lead to some species becoming **endangered**.

pollution

The introduction into the environment
of harmful substances.

raw materials

The basic materials from which a product
is made, for example wood or metal.

renewable energy

Energy from a source that does not
run out. Wind power and **solar power** are
examples of renewable energy. They are
sometimes simply known as "renewables."

solar power

Energy that comes from the light and heat
of the Sun.

sustainable

Able to exist while causing little or no damage
to the environment.

synthetic

Made from artificial chemicals, not natural materials.

toxic

Poisonous.

zero carbon

Causing no **carbon emissions**. Another term
for zero carbon is zero emissions.

INDEX

A

B

C

F

fossil fuels, 11, 23-27, 38, 97

fracking, 26-27

free range (animals), 128

freecycling, 57

freezers, 44, 152

fruit, 105, 143-144, 189

G

gardening, 144, 189, 191

gasoline, 155-156

geothermal power, 33

glasses, 94

glitter, 108

green hydrogen, 33

greenhouse effect, 25

greenhouse gases, 24-25, 60, 135

P

sustainable forests, 60, 192
sustainable products, 115, 192
synthetic fabrics, 65-67, 108-111

T

tidal power, 35
toilets, 169, 178
trains, 154-155, 157, 159-160
travel, 11, 24, 46, 132-133, 135, 154-160
trees, 59-60, 181, 185, 191
tuna, 131

U

upcycling, 78

V

vegetables, 105, 141, 143-144, 152, 189

Additional editing by
Susan Meredith, Kristie Pickersgill
and Kirsty Tizzard

Americanization by Carrie Armstrong

Usborne Publishing Ltd., Usborne House,
83-85 Saffron Hill, London EC1N 8RT, England.

usborne.com

Printed in China. First published in 2020. Copyright © 2020 Usborne
Publishing Ltd. AE. Library of Congress Control Number: 2020935218